BookCaps Presents

Genetics

In Plain and Simple English

By Aimee Ogden

■BOOKCAPS

BookCaps™ Study Guides
www.bookcaps.com

© 2014. All Rights Reserved.

Cover Image © adimas - Fotolia.com

Table of Contents

INTRODUCTION 3

CHAPTER 1: A PUZZLE OF PEAS 7
   GREGOR MENDEL, NEUROTIC MONK GENIUS 8
   PEA PORNOGRAPHY 10

CHAPTER 2: PREDICTIVE POWERS 16
   OBEY INDEPENDENT ASSORTMENT – IT'S THE LAW 16
   BY THE NUMBERS 19

CHAPTER 3: THE DISCOVERY OF DNA 24
   KILLING MICE (BUT, LIKE, FOR SCIENCE) 24
   NO IMAGINE DRAGONS REFERENCES, PLEASE 27

CHAPTER 4: THE MOLECULE OF LIFE 31
   DOUBLE HELIX – DOUBLE TROUBLE 31
   A MIGHTY MOLECULE 34

CHAPTER 5: THE CENTRAL DOGMA 40
   A BIG BAG OF GOO 40
   DON'T SHOOT THE MESSENGER 45

CHAPTER 6: COPYING THE FLOPPY DISK OF LIFE 58
   MITOSIS 58
   MEIOSIS 65

CHAPTER 7: GENETIC HEALTH GOES "BOINK" 72
   WHERE'S THE LASER VISION GENE? 72
   MY, OH, MEIOSIS 78

CHAPTER 8: EXCEPTIONS THAT PROVE THE RULE 82
   IT'S NEVER THAT SIMPLE (EXCEPT WHEN IT IS) 82
   INCOMPLETE DOMINANCE 83
   CODOMINANCE 84
   PLEIOTROPY AND EPISTASIS (ARE NOT TWO WORDS I JUST MADE UP) 86

POLYGENIC TRAITS 87

Introduction

Biology is boring.

At least, that's what you remember from your school days. There were endless lists of things to memorize; parades of dead white guys who discovered how to make the world's grouchiest face for the camera as well as figuring out how genes work or how organisms evolve; and hours of peering through a microscope at – what the heck *is* that? You're looking at your own eye somehow. How did you even screw up *that badly*?

Here's the thing, though: biology shouldn't be an interminable slog through stupid acronyms and dissections where you end up wearing more of the specimen than you're identifying. Biology is the study of the frankly amazing voyage our genes have made over the last three billion years to carry us from the dawn of life on earth to the endless forms most beautiful we see today (and in some cases, the endless forms most freaky, most grotesque, and most likely to inspire screaming when encountered on the bathroom wall). Through the half-assed engineering process known as evolution, we have the world's most complex supercomputer installed in every single one of our cells – and like every computer ever invented, DNA is just as hacky and likely to crash and burn.

If you're interested in knowing more about the weird things going on inside your cells right now – but without all the tedium – then keep reading. We're going to start with an experiment to get you intimately acquainted with your DNA. Do you have some vodka in the freezer? Good. (Unless you're under the drinking age – then skip straight to Chapter One. Alcohol is for losers, kids! Just say no!)

All right, where were we? Vodka. Make sure you have an ounce (or more, just in case you have some more experiments planned) chilled ice-cold in the freezer. You're also going to need pineapple juice (just a drop), or contact lens solution if you must, liquid dish soap, and a pinch of salt.

Oh, and about a quarter of a shot glass filled with your spit. Get salivating! While you're working on your Gaston-esque levels of expectoration, be sure to scrape your tongue against your teeth. Don't draw blood or anything – you just want to be sure that you're grating off that top layer of cells. (Don't worry, they're going to die soon enough anyway; you aren't going to miss them.)

When you've got your 1/4 glass of cell-filled drool, go ahead and add a single drop of the dish detergent. Ever pour a drop or two of dishwasher liquid into a greasy pan and watched the fat droplets scurry away in terror? The outside coverings of your cells – and of the nuclei inside, where your DNA is stored – are made of the same fatty material, and feel about the same way about dish soap. The soap breaks up those coverings so that the contents of your cells are dumped out into the surrounding liquid.

But as you may (or may not) recall from those biology classes way back when, your DNA is packaged tightly up as chromosomes. There are two meters of DNA in each of your cells – that's almost a Wookiee's worth, for those of you who don't play by metric system rules – but when it's packaged up tightly in chromosome form, it's wound up so snugly that it fits inside each of your microscopic cells. It's a bunch of tiny protein molecules that do the packaging, and we don't want them to come along for the ride if we're trying to isolate your DNA. Add a drop of pineapple juice or contact lens solution – these liquids contain enzymes called "protease" (that's PRO-tee-aze) that rip proteins into shreds.

So by now, you have some loose DNA floating in your shot glass – but it's still dissolved in the water of your saliva. Add a pinch of salt – that helps block the DNA from interacting with the water molecules, which will help un-dissolve the stuff. Finally, tilt the shot glass to the side and slowly – SLOWLY – pour the ice-cold vodka in. If you pour too fast, you'll basically rip the DNA apart with how quickly the liquids are mixing. Be kind to your DNA – pour gently.

As you pour, you're going to see a white, slimy substance beginning to appear where the alcohol and saliva mix. That, my friend, is your genetic blueprint. And it looks like snot!

Continue pouring until your shot glass is full. Admire the snotty majesty of your DNA (and whatever other cellular debris is clogged up in its two-meter-long tendrils). Throw back your shot. How does the operating system of your life taste?

Good. Now you're ready to learn about genetics.

Chapter 1: A Puzzle of Peas

Gregor Mendel, Neurotic Monk Genius

The foundation of modern genetic theory was laid down by an Austrian monk who was had such anxiety problems that he couldn't make it through a teaching certification test without having a nervous breakdown. His discoveries were published in 1866, but wouldn't get any traction with the scientific community until almost thirty years later – because they were written up in a German-language journal by an Austrian monk who had such anxiety problems that he couldn't make it through a teaching certification test without a nervous breakdown.

Gregor Mendel wasn't the first scientific thinker to grapple with the problem of how traits are handed down from parents to children. Scientists had been aware for a long time that, somehow, characteristics are passed down through the generations – and not just scientists, either. Farmers and herders were aware enough of heredity to incorporate the basic idea into their breeding programs; for that matter, Thogg the Neanderthal probably knew enough to get suspicious that Thogg Junior looked more like the Cro-Magnon milkman than he did his own dad. But what no one could figure out was how, exactly, this transmission happened.

Early theories included the "homunculus" idea – basically, that Dad's sperm contained a miniature, perfectly formed human (the homunculus) in every drop, and that Mom just provided the ideal growing conditions necessary to get the little critter big enough to survive in the outside world. But that didn't explain why kids sometimes shared traits with both parents, not just their father – if Pops was providing all the material to actually make that baby, how did the incubator contribute a hair color or ear shape?

Another concept was "blending theory". Mom has traits, Dad has traits, and these get mixed together to form Kiddo's traits. That explains everything … except why children aren't always a bluish-brown-eyed, brownish-black-haired, intersexed cross between their parents. If traits get blended together through generations, why do we still have distinct features like pointed tongues or dimples or red hair instead of a mishmash of whatever our parents had?

Or maybe what happened is that you got some traits from one parent, and some from the other: you get your dad's hair, but your mom's eyes; your father's tuba-playing skills, and your mother's pathological disdain for pineapple. This seemed a little closer to the mark, and wouldn't result in any bizarre halfway-in-the-middle mixed traits, but no one had really figured out a way to prove this as an explanation, either. (Spoiler alert: this is because it's not actually the explanation.)

Despite his previous academic trauma, Mendel was a pretty smart guy, and set about attacking the problem of heredity in a twofold way: with gardening, and math. The abbey he lived at after departing the academic world grew its own food, and he took over part of that garden in order to plant peas. Thousands and thousands of peas.

Pea Pornography

As far as genetics experiments go, peas are a pretty good option. For one thing, they have a large number of either-or traits (green or yellow seeds, pink or white flowers, short or tall heights) – that makes for much easier analysis than complex traits like human height or skin color, which fall along a broad spectrum. For another, it's a lot easier to control the breeding of peas than it is to do with mice, which will shamelessly mate with their siblings if you don't separate the litters fast enough. All you have to do is rip the male parts of the flowers out (ouch) before they start pollinating themselves. (Yes, in the plant world, you can get yourself pregnant by masturbating.) And finally, and probably most on point for Mendel, peas are fairly easy to grow, harvest and store; they're cheap; and if you have to grow your own food anyway, you might as well choose something you don't mind eating for the next forever.

Mendel planted "true-breeding" strains of pea plants (plants whose offspring always had the same version of a particular trait – always wrinkled seeds rather than smooth, always pink flowers rather than white, and so on), and then bred individuals from different true-breeding varieties together. He did this for a lot of different varieties, because, aside from all the praying, monks have a lot of time on their hands; but we'll focus on the example of plants that produce green peas and plants that produce yellow seeds.

Mendel crossed a green-pea plant with a yellow-pea plant, and sat back to see what would happen. If the blending hypothesis were true, then he should get yellowish-green peas; or if the baby plants got a mix of traits from Pea Mom or Pea Dad, then he should see approximately half yellow-pea plants and half green-pea plants.

Instead, Mendel was baffled to find that exactly 100% of the new plants had yellow peas. The green pea trait had apparently been completely wiped out in this new generation, which threw any possible "blending" of parental traits right out the window. Surely the yellow pea plants hadn't contained tiny pea homunculi in each little piece of pollen? (In case you didn't know, pollen contains plant sperm cells. Try not to think about what that means next spring during allergy season.)

He was at a scientific dead end, but Mendel decided not to end the experiment there. Instead, he let the plants fertilize themselves to produce a second generation. This time, there *were* green-pea plants among the yellow ones – even though there hadn't been any green-pea parents! Not only that, but the ratio of yellow-pea plants to green-pea plants in this generation was almost exactly 3:1 – and the same was true in the second generation of every other such cross Mendel attempted. Flower color, seed texture, plant height – every single time, there was 100% representation from one trait in the first generation, and a near-perfect 75%/25% split in the second. (A suspiciously near perfect split, in fact, leading some science historians to speculate that the pious Mendel may have not-so-piously fudged his data a bit.)

So what was going on? Mendel hit the books, and finally hit upon an explanation that made sense with the repeatable mathematical patterns that kept turning up in his garden experiments. What if, instead of a single "hereditable factor" to control every trait in a living thing, there were two such factors – one from each parent? And what if the different versions of these hereditable factors could sometimes override one another – so that if you had a different version from each parent, you'd only see the version that "won"?

So, Mendel, reasoned, the true-breeding strains he'd started with all had two identical versions of the pea-color hereditable factor. The green-pea plants only had two green-pea factors; the yellow-pea plants each had two yellow-pea factors. When a green-pea plant and a yellow-pea plant were allowed to get whatever the plant version of funky is, the offspring got one green factor from the green parent, and one yellow factor from the yellow factor. The yellow factor is dominant to the green one, so only yellow peas are made by the new baby plant – the green factor is wimpy, or "recessive", and although it's still there, it doesn't get to do anything.

And then things get a little more complicated in generation 2. The half-green, half-yellow hybrid plants have a 50-50 chance of passing on each color to their own offspring. If the new plant gets two green factors, then it gets green peas. One yellow factor and one green factor, and the yellow one overpowers the other – yellow peas are all that are produced. And two yellow factors of course means yellow peas, too.

Wait, hang on – doesn't that mean there's a 33% chance of getting green peas, not a 25% chance? Well, no; because you can get a green factor from the sperm cell and a yellow factor from the egg cell, or you can get a yellow factor from the sperm cell and a green one from the egg cell. Those are technically two entirely different possibilities, and we have to count them separately – so there's only one way to get a green-pea plant in generation 2, and three ways to get a yellow-pea plant. Don't blame me; blame statistics.

So Mendel, quite proud of himself and his pea plants, wrote up his results and sent them off into the wide world of academia to see what other scientists had to say. Unfortunately for Mendel, a lot of those scientists' mothers had never told them that they shouldn't say anything at all if they didn't have anything nice to say … and the rest were silent because their mothers had. Blending theory was "obviously" the right theory, and any evidence to the contrary had to be wrong. So Mendel's work was published in an obscure German journal, and he died in the same obscurity in which he had lived (which was probably all right for him and his history of nervous breakdowns). No one cared about his weird ideas about hereditable factors, and frankly, the amount of pea sex he was facilitating was a little creepy. The world was more than ready to write off Gregor Mendel for good …

… until the turn of the 20th century, when scientists were finally ready, after decades of being unable to come up with a plausible explanation, to reject blending theory. A pair of scientists, Carl Correns and Hugo de Vries, independently arrived at the same conclusions as Mendel did – only to realize that their Earthshattering Discovery had already been long since been discovered, unraveled, and published by a dead Austrian monk. (Nowadays, the academic mantra is "publish or perish", but Mendel had decided to go both routes.)

To add insult to injury, Mendel's experiments and conclusions were even more elegant than the ones these scientists came up with forty years later. One thing which we can wholeheartedly credit de Vries with, though, is the coining of a new term to describe Mendel's laws. Instead of the mouthful "hereditable factor", de Vries called the material passed down from parent to offspring "pangenes" – or, as we call them today, simply "genes".

Chapter 2: Predictive Powers

Obey Independent Assortment – It's the Law

By obsessively breeding together peas, Mendel figured out a few basic rules of heredity almost fifty years before any other scientists got on the genetics bandwagon. These three rules are fundamental to the science of genetics – especially when scientists work to figure out just how all the exceptions to those rules are able to happen. Here's the breakdown of Mendelian genetics:

1. There are different versions of genes, and that's what causes differences in traits. For example, there's an "eye color gene" (actually, that situation is a little more complicated than you might have learned in high school biology, but more on that later), but then there are blue and brown and green and gray versions of that gene. These versions are called "alleles".

2. You have two copies of each gene, because you get just one from each parent. Everybody (and everypea) has a matched set of each gene: one that they got from Mom, and one that they got from Dad. Both copies might have the same version, or allele; or they might be two different alleles. If and when everybody has their own kids (or sprouts) someday, everybody only gets to pass on one of their two alleles. If this didn't happen, the next generation would have four copies of each gene, and the one after that would have eight, and so on; and turning everybody's genetics into a game of 2048 is frankly not a good idea. You get one copy of each gene from each parent; you pass on one randomly chosen copy of each gene to your kids. That's the deal, and it's called the Law of Segregation.

3. If you've got two different alleles, one allele dominates the other one. As in the pea-color example, a pea with two different seed color alleles – one green, one yellow) – will only show the dominant trait of yellow seeds. The recessive allele just goes into hiding.

4. The allele you get from your parent for one gene doesn't affect the allele you get for the other gene. Your dad has both a yellow and a green allele for pea color. (You're a pea plant, right? Good.) He also has both a pink and white allele for flower color. The chances that he passes on the yellow allele to you has nothing to do with the chances that he passes on the pink allele to you, and vice versa. (This rule is only partway true, for reasons we'll get into when we start talking about what "genes" actually look like and where to find them. For the traits Mendel happened to study in pea plants, it happens to work out 100% every time, which is another reason to suspect he might have fudged his data; if he accidentally chose a trait that didn't follow this rule, he probably just pretended that experiment never existed.)

Now, one of the keys to doing good science is being able to make predictions that are carried out by an actual experiment, and Mendel's rules of genetics let him do just that. He could predict the ratios of traits in the offspring of any pea plants he crossed – and we can do the same in animal breeding, and even in human genetics.

By the Numbers

If you've had kids, you may have been offered the chance for a genetic screening beforehand. It may seem that having the doctor draw you and your mate's blood to try to predict traits in your offspring isn't much more than modern-day dark magic – but the hospital lab can take a peek at what alleles are floating around in your cell, and use that information to figure out what your theoretical kid's chances are for a lot of severe genetic diseases. A pretty hefty number of nasty issues can be tidily predicted based on Mendel's rules for inheritance: cystic fibrosis, for example. If you get tested and find you have one good allele and one bad one for this disease, you are what we in the business call being a "carrier": you don't have this disease yourself, because your good allele is dominant over the crappy one, but you're toting the allele associated with it around all day. (This is also known more generally as being "heterozygous"; *hetero-* is a Greek root meaning different, as in, you have two different alleles for this gene. "Carrier" is a word used strictly to mean toting around an allele that could inflict some harm on your offspring; "heterozygous" is the more general use case of having two different alleles, good or bad.)

In the case that you do turn out to be a carrier, it all depends on what's in your partner's DNA. If they're free and clear of the bad allele, then your kid has zero chance of getting cystic fibrosis – they only get one version of each gene from each parent, so even if they get a bad copy from you, then your partner is picking up your genetic slack with a good copy. Go ahead and get boning! (Two identical alleles for any given gene is called being "homozygous"; just like *hetero-* means different, *homo-* means same.)

It's pretty useful to be able to make this kind of prediction based on your knowledge of someone's genes. In school, you probably learned about using Punnett Squares in order to make this kind of prediction, but we're going to skip Punnett Squares, because we are grown-ass adults and we can use basic math (by which I mean, we can use calculators). All you need to know is what particular collection of alleles each parent has – what scientists call the "genotype".

Let's represent the healthy version of the cystic fibrosis gene with the letter F, and the worthless garbage version with the letter f – capital letters are typically used for dominant versions, and small letters for recessive ones. So the genotype for a parent who has no bad version of the gene would be FF; someone who's a carrier would be Ff. A kid who ends up with the disease would, of course, be ff (ucked). What are the chances that two carriers would have a child with cystic fibrosis?

To have cystic fibrosis, the kid has to get an 'f' allele from both parents, so all we have to do is figure out the chances of each parenting handing down that type of allele. For a carrier (Ff), the odds are of course going to be fifty-fifty. Thanks to that Independent Assortment business, half of the time, they'll pass down an F, and half the time, an f.

To figure out the odds of getting that f from both parents, you just multiply the odds of getting it from each parent. A 50% (or 0.5) chance of getting an f from dad times a 50% (or 0.5) chance from mom gives a 25% (or 0.25) chance of having a kid with the world's worst mutant power: super-sticky mucus.

What are the odds of getting a kid who's healthy? Well, you can do this the hard way, and calculate the chances of each of the genotypes that give a healthy kid (FF, Ff, and Ff again – don't forget, getting the F from Mom and the f from Dad is a completely different possibility from getting the f from Mom and the F from Dad. Again, this is statistics' fault, not mine.) If you prefer things less complicated, you can just subtract the percentage chance of having a diseased kid from 100. So if you've got a 25% chance of having a kid with cystic fibrosis, you have a 75% chance of without it.

So science can help you figure out, based on your genotype, or your mate's, what your chances are of having a particular trait show up in your kids. Science can't, of course, help you decide what to do with that knowledge. That's what Magic Eight Balls are for.

Sometimes, you don't need a scientific test to know what your genotype is – you can tell based on your own traits what your genes must be secretly doing inside your cells. If you have dry earwax, you know you have two recessive alleles, because the gene for wet earwax is dominant. If you have wet earwax, you might either have two dominant alleles (EE), or you could have one of each allele (Ee) – there's no way to know based on the trait alone, because two different genotypes give the same result.

Unless you have a kid with dry earwax! In that case, the child has the genotype ee – two recessive alleles, without a dominant one to cancel them out. That means both you and your partner must have at least one little-e allele – that's the only way for the kid to get two copies of it.

And what if you and your partner both have dry earwax, but your child shows up with the wet variety? Well … that means it's time to call Jerry Springer. Sorry.

So Mendel's rules provide the lay of the land for us to make inferences and predictions about our offspring and about what's going on inside our cells. But how does it all work? Mendel only guessed that these nebulous "hereditable factors" were involved; he had no idea what they were made of or how they got from parent to child. It took several decades, complicated lab equipment, some radiation exposure, and a great deal of scientific in-fighting and back-stabbing before anyone could point to what, exactly, carried traits from one generation to the next.

Chapter 3: The discovery of DNA

Killing Mice (But, Like, for Science)

Once Correns and de Vries had established that Mendel's theories were true and that hereditable factors, or genes, carried hereditary information, the next step for scientists was clear: someone had to figure out where and how, exactly, this information was stored.

Imagine you're a scientist at the turn of the 20th century; your scientific equipment includes some glassware, syringes, and a microscope roughly equal in power to what you can find in a middle school classroom today. There was no way at all to take a peek at the actual molecules floating around inside cells – so how could anyone prove what was happening at the molecular level? How could you ever say for sure that there was some miniature physical substance in which these genes – these hereditable factors – were contained?

Well, as with so many things in science, you start by killing some mice. Frederick Griffiths was an English microbiologist working for the National Pathology Laboratory, and he designed an experiment that would show that there was a physical material that transmitted hereditary information while simultaneously acting as an elaborate form of pest control.

Griffith's experiment relied on his use of two different forms of pneumococcus bacteria – a microorganism that causes pneumonia. One strain (the smooth, or "S" strain) was deadly; injecting mice with that strain gave you a big pile of dead mice. The other strain (rough, or "R") was harmless: the worst ill effect that a mouse injected with R strain bacteria would suffer was the needle-poke itself.

Griffiths tried heating a vial of the deadly S strain up to a temperature hot enough to kill all the bacteria, and then injected that mixture into the mice. Unsurprisingly, the mice were fine – dead bacteria aren't any better at infecting mice than dead cats are at hunting them.

But his next step was to take a vial of those heat-killed S strain bacteria and mix their tiny little bacteria corpses in with a healthy vial of perfectly harmless R strain bacteria. When he injected that mixture into mice – they started dropping dead. Wait, what?

How could it be that the harmless R strain bacteria or the dead S strain bacteria had killed any mice at all? Something must have happened to the R strain bacteria to give them the same powers of pathology that the S strain had had before their untimely bacterial demise. Griffiths reasoned that the S strain bacteria had contained a physical substance: once they had been destroyed by the heat and mixed with the R strain bacteria, those R strain bacteria had ransacked their cellular corpses to take up that material. Once the material was inside the R strain bacteria, they were able to start acting like S strain bacteria – they had the traits created by the S strain's hereditary material. (Do not try this at home. You are not a bacteria, and eating your friends will not get you their powers – it will get you a jail sentence in a tiny, padded cell.)

But Griffiths had only showed that there was a physical hereditable material … not what that material was. It would be more than twenty years before anyone was able to find a way to put the next piece of the puzzle in place.

No Imagine Dragons References, Please

Despite the lack of microscopic magnification available, scientists could use other biochemical methods to figure out what sorts of biological molecules were floating in the tiny goo-filled bags that are cells. Of that long list, there were two main candidates that stuck out as potential information carriers: DNA, and proteins.

DNA and proteins are both on the short list because they form long chains – which is what you'd expect would be necessary to carry the sort of complex information that would go into creating an organism's traits. Each one also has a few specific subunits that make up those long chains: for DNA, those subunits are called nucleotides, and for proteins, they're amino acids.

And those subunits are precisely why scientists believed, until the 1950s, that proteins had to be the thing that carried our genes: proteins have about twenty different amino acids to use to string together their chains, while DNA has … four. Imagine trying to describe your hair color using only four letters – let alone provide enough information to actually recreate that hair color!

But going about proving that protein was the material of heredity was tricky. It's not as if you can just take all the proteins out of an organism and see if it still passes on its traits to its offspring – despite the simplified diagrams textbooks show, cells are complicated messes cram-packed with protein, as well as a lot of other gunk, from top to bottom. So if cells don't work, what's the next best thing? Viruses.

Viruses don't really count as "living things". They're a wad of DNA (or RNA) wrapped up in a protein case, and that's about it. They're pretty much tiny machines that exist solely for the purpose of making more copies of themselves. They latch on to the outside of living cells, dump hereditary material into those cells, and turn them into miniature virus factories. But which piece is it that provides the instructions on how to make viruses: the DNA wad, or the protein wrapper?

Scientist Alfred Hershey and his lab assistant Martha Chase devised a way to figure out what was going on, and that way involved radioactivity and blenders: no glow-in-the-dark margaritas, sadly, but just an incredibly elegant experimental design.

As it happens, sulfur is an element that is found in proteins (certain amino acids use it as a building block) but never in DNA. On the other hand, phosphorus is a major component of DNA (every single one of those four nucleotides contains it), but you won't find it in any amino acids. So Hershey and Chase (well, let's be real, the lab assistant was probably the one doing all the heavy lifting) created two different virus strains: one that had radioactive sulfur in its proteins, but "regular" DNA; and one that had radioactive phosphorus in its DNA, but "regular" protein.

In two separate containers, they combined one of their two experimental viruses with some tasty bacteria to infect. They then waited just long enough for the viruses to shoot their hereditary material into the bacteria – and then put the whole mix in a blender.

Hang on: why would they ruin a perfectly good smoothie-maker with radioactivity and bacteria slime? Well, at high enough speeds, viruses that have been latched on to the outside of bacteria (for infection purposes) will pop right off. And if you let your blended bacteria colada settle for a few moments, the fact that bacteria are huge (relatively) and viruses are tiny will work to your advantage: the bigger bacteria will settle down to the bottom of the container, and the viruses will float freely in the liquid.

Bacteria infected with viral hereditary material settle on the bottom, the unwanted, non-hereditary virus bits floating on top – see where this is going? Now all you need is a Geiger counter to check out where your radioactivity is hanging out. In the experiment that used radioactive sulfur (the stuff in proteins, remember), the gloppy bacterial sediment at the bottom of the flask didn't ping the Geiger counter – but it rang its beepiest alarm for the virus-y liquid on top. And in the experiment with radioactive phosphorus labeling the DNA, the Geiger screamed its mechanical head off at the bacterial goo, but couldn't have cared less for the fluid full of depleted virus.

If DNA was getting into bacterial cells when they were infected by viruses, but protein was not, there was no way protein could be the hereditary material. It had to be DNA, four nucleotides notwithstanding. But how on earth could any kind of meaningful message be sent using only four letters?

Chapter 4: The Molecule of Life

Double Helix – Double Trouble

By the 1950s, it had been established that DNA was the stuff of inheritance, but nobody had any clue what, exactly, that "stuff" looked like. At last, though, scientific technology was beginning to catch up to what scientists wanted to use it for: by crystallizing a solid chunk of DNA and shooting X-rays at it, then measuring where those X-rays ended up once they'd bounced off the DNA, researchers could start to figure out how DNA was shaped. Think of it this way: imagine there's a dark room with a big shape in the middle of it. If you throw a rubber ball at that shape and it bounces straight back to you, you know the shape has a flat face; if it bounces up or down, you know the shape is curved vertically, and if it goes left or right, the curve is horizontal. By walking around the shape and throwing the ball continually, you can get a vague idea of the whole structure. That's pretty much what scientists were doing, except instead of a ball, there were hazardous levels of radiation involved.

Two researchers particularly interested in being at the forefront of any DNA-structure-related discovery were James Watson and Francis Crick. Watson, an American, and Crick, an Englishman, were absolutely whatever the scientific equivalent of "thirsty" was – they kept on trying to dig into DNA's mysteries even after being told by their boss to work on something they had a chance of actually figuring out.

But Watson had attended a talk given by another scientist, Rosalind Franklin, who had gotten some impressively good X-ray images of DNA – images that only made Watson even more interested in what was going on inside those little tiny DNA molecules. Franklin had figured out that DNA had two different forms, based on the moisture level in its surroundings, and she thought that both of those forms were somehow shaped like a spiral, or "helix". She wanted to nail down proof that this was true (she could show that one of the two DNA forms was helical, but not that they both were, based on her results so far).

Too bad for her, her boss, Maurice Wilkins, was impatient with her insistence on scientific surety, and showed her data to those nice young boys Watson and Crick. Based on the X-ray diffraction images Franklin had collected, Watson and Crick were able to figure out that DNA wasn't just a helix – it was a *double* helix. Two long strands of DNA were bonded together, like the twin railings of a spiral staircase. Watson and Crick rushed to publish a one-page paper that all but contained the phrase "IN YOUR FACE, ROSALIND" in every paragraph. They'd scooped her and undercut all her hard work – a fact for which Crick felt bad later, calling the manner they'd treated her "patronizing" at best. Watson never hemmed or hawed over the situation, though – in his view, Franklin was an ugly loner who didn't belong in the lab and who deserved whatever she got (and what she got, to be clear, was having one of the biggest scientific discoveries of all time unceremoniously yanked out from under her. Gosh, it's hard to imagine why she ever might have had a chip on her shoulder.).

Ten years later, Watson and Crick and Wilkins won the Nobel Prize for their DNA work. The prize is only allowed to be given to three recipients, and it's only allowed to be given to the living – and Rosalind Franklin had passed away four years before. (She died of cancer, possibly due to doing all the radiation-exposure grunt work only for these two young turks to swoop in and get all the glory.) Maybe she would have been the third honoree instead of Wilkins if she'd still been around, maybe not – but optimistic might-have-beens don't really make up for how she was treated during her life.

A Mighty Molecule

Watson and Crick (and Franklin) established that DNA is shaped like a double helix – but what does that mean, really? There are two DNA strings all twisted together – but what does that look like? And how were they able to make the logical leap from Franklin's data to surmise the structure of DNA?

Let's start by reiterating something that came up before: DNA is a polymer, a long chain of small pieces connected in a row. (Or, to be quite accurate, each molecule of DNA is made up of *two* chains.) Each link in the chain – each basic building block – is called a "nucleotide", and DNA contains four different types of nucleotide.

But really, those basic building blocks have even more basic components. All four models of nucleotide are made of three parts: a phosphate group, a sugar called deoxyribose, and a "nitrogenous base" (don't panic; that just means it contains nitrogen. Scientists never met a four-syllable word they didn't like). The phosphate group and the sugar are exactly the same no matter which of the four nucleotides you pull out of a hat, but the base is different in each of these four. It's this one small difference in shape that allows DNA to carry information – as important as the difference between the letter "l" and the letter "t" that keeps you from saying "tit" instead of "lit".

Two of those bases are called cytosine and thymine (henceforth called C and T), and their structure contains one small ring of nitrogens and carbons. The other two bases, adenine and guanine (A and G), are made up of two of these rings – they're about twice as big as C or T.

Watson and Crick were aware that DNA was made of nucleotides and that those nucleotides were made up of a phosphate, a sugar, and a base apiece. They knew from Franklin's data that a spiral, or helix, was involved somewhere. And the last piece of the puzzle was from data published by a scientist named Chargaff, who'd figured something very strange about DNA: if you took a sample of the stuff and measured how much of it was A's, how much was C's, and so on, you'd always – always! – find that the amount of A was exactly equal to the amount of T; and the amount of C was equal to the amount of G. It didn't matter if the sample came from a human, a hippopotamus, or a hyacinth – the ratios were always the same.

The same number of A's (a big base) as T's (a small base), and the same number of G's (big) as C's (small) – what could explain why these bases seemed to be marching in lockstep? Maybe if they were, in fact, locked together. Franklin had figured out that DNA's shape involved a spiral – but what if you had two spirals stuck together, like a twisted ladder? An A on one side would be stuck to a T in the other to form a rung; and the same was true for C's and G's:

Figure 1: Two nucleotide buddies.

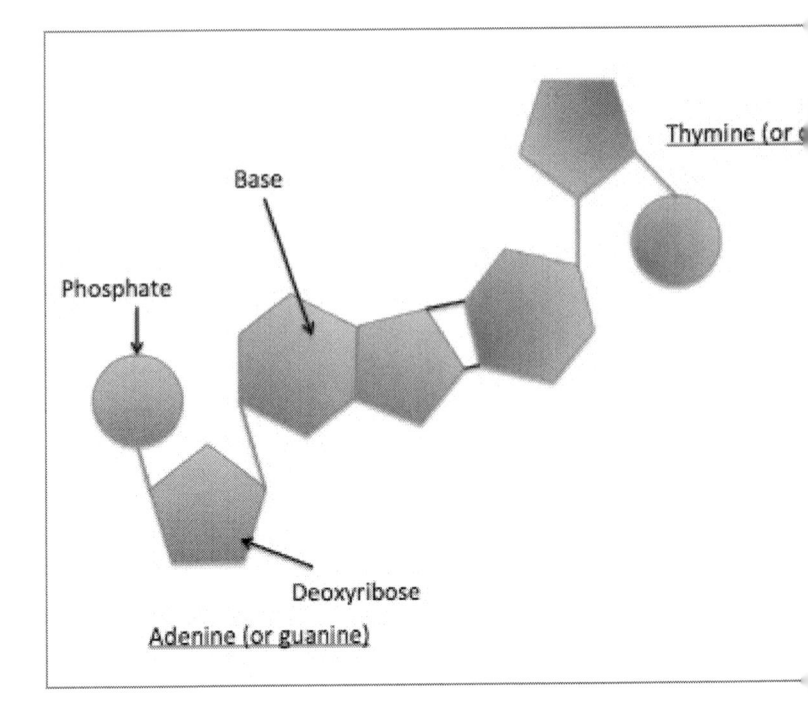

If a "big" base was always stuck to a "small" base, the ladder would always have the same distance from side to side – whereas rungs would get weirdly wide in a spot where you stuck two A's together, or pinched inward if you tried to bond a T to another T. Imagine a ladder whose two sides wobble inward and outward, and you can guess that besides explaining Chargaff's weird ratios, a structure like this is also a lot more stable than the alternative.

So the bases form the creamy nougat center of the DNA helix, with A's happening to have just the right shape to fit together with T's, and C's with G's. The sugars and phosphates, then, form the outside, alternating one after the other and chaining together with the phosphate of one nucleotide clinging on to the sugar of the next one, and so on:

Figure 2: Not picture - the other several million nucleotia the strand.

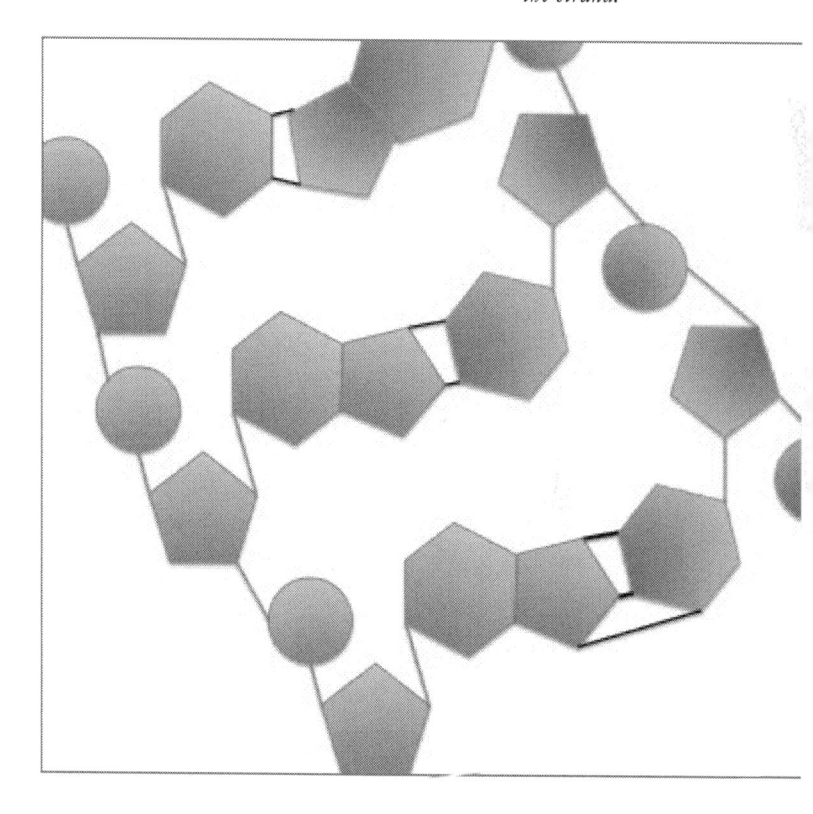

If you take a close look at the above figure, you may notice something about the two strands of DNA. (Or you may not, depending on what you did with the bottle of vodka after you took your DNA shot.) The blue strand is pointed "upwards", with the sugar/base of each nucleotide on top and the deoxyribose sugar on the bottom. The red strand, on the other hand, is pointed "down", with all the pieces of each nucleotide in the opposite orientation. The two strands of DNA are indeed "antiparallel" – they run in opposite directions, so that if you followed a piece of DNA all the way to the tip, you'd find a sugar at the very end of one strand, and a phosphate at the very end of the other. Do you care yet? Not really – but file this information away for later, when we talk about how DNA makes copies of itself.

But now that we've zoomed in to DNA about as far as it's possible to zoom, let's stay on the molecular level for a little while. We know that these little spiraling strands of A's, C's, T's and G's somehow store enough information to create the traits that Mendel observed in his garden ... but how does ACCTTAGGAC spell out "pink flowers"? What's the mechanism to turn the language of DNA into the language of distinguishing characteristics and features we see around us?

Time to review some cell biology.

Chapter 5: The Central Dogma

A Big Bag of Goo

Did you study diagrams of the cell in your past biology classes? If so, they probably looked something like this (except maybe a little bit less shaped like a whale with elephantiasis):

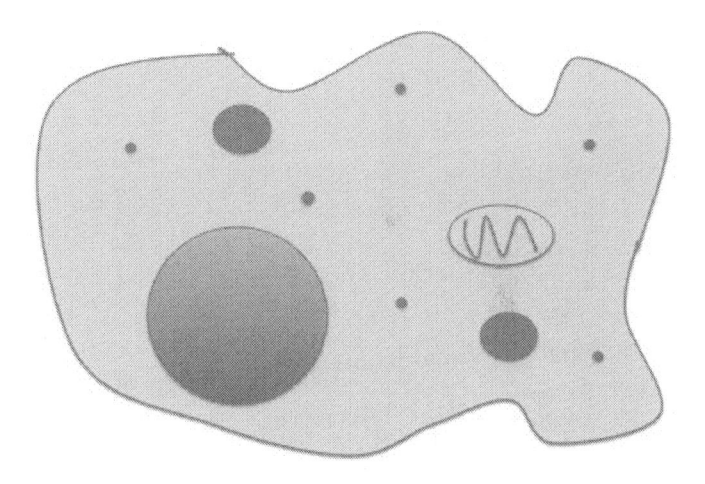

Figure 3: Nailed it.

Most representations of cells look more or less like this: a big bag of cellular slime with a few odds and ends floating around inside. This is because it's pretty hard to illustrate the reality: the fact is, your cells are jam-packed with structures and molecules from stem to stern. No one puts accurate pictures of the insides of cells in textbooks because no one wants to make biology students cry (except maybe the Advanced Placement people – those guys are scary).

Your cells are packed with sugars and fats being broken down for energy, with protein spikes that keep the whole thing from collapsing like a microscopic whoopee cushion, with little tiny pieces of machinery to break down waste and get it the heck out of there. There are miniature structures, called organelles, that do all the little tasks cells need to do to stay alive, and they're crammed into the cell from stem to stern.

And the most important organelle is the nucleus of the cell, that big blue bubble on the diagram above – that's where your DNA is hanging out, and it's no exception to the hustle-and-bustle going on everywhere else. During most of your cell's life, the DNA is unraveled in there and flung in every direction for little machine readers to browse around and find the bits they're looking for. "What, Bob ate a pint of ice cream for dinner again? Okay, let's find the piece of instructions to make the enzyme to break down all that dairy. He's been hanging out in the sun all day? All right, we're gonna need the bit that tells us how to make melanin!"

DNA on its own is sort of a limp rag – a limp rag that happens to contain the entire key to your existence. DNA is really smart, but also really helpless.

Proteins, on the other hand, are dumb, but their much more complex structures (remember how there are 20 different amino acids available to make proteins compared to only 4 different nucleotides) allow them to do a lot more jobs in your cell. They break down foods, make hormones and transmitters, provide structure and shape to cells, and move things around. Proteins make stuff happen. Proteins are the reason we have different traits – pink flowers and blue eyes and green seeds and cystic fibrosis. And it's the information on how to make proteins – thousands and thousands of different proteins – that's stored in your DNA. DNA is an instruction manual on how to make an organism, and proteins are the machinery that does the making.

As it happens, proteins are also a lot smaller than the average DNA molecule (or chromosome). You have 46 chromosomes in each of your cells, and the *smallest* one is 50 million nucleotides long. The biggest is a chain of almost 250 million nucleotides, and if you think that behemoth is doing anything but lying around the nucleus waiting to be brought the cellular equivalent of bon-bons, think again. Unfortunately, while all the information is stuck in the nucleus with DNA, the actual protein-making machinery – a type of organelle called "ribosomes" – are located outside the nucleus, floating elsewhere in the cellular schmutz, and they're too big to get into the nucleus just like the DNA is too big to get out. Information in point A – machinery in point B – how the heck to get from one to the other?

DNA relies on a small army of microscopic molecules to read it, transcribe copies, and carry them out into the cell to find a ribosome to do something with all that information. Except for the brief periods when the cell is copying itself, there is a continuous process going on in which messenger molecules scribble down little DNA notes and take them out to be handled. These messengers are made up of a type of molecule called RNA – DNA's more athletic cousin. If the name seems familiar, it's because RNA is a very similar type of molecule to DNA, with a few small differences:

1. Both are made up of nucleotides, but RNA's nucleotides are made using a different sugar than DNA's are. DNA's sugar is deoxyribose

– RNA's is just plain ribose; hence the difference in names (DNA is short for deoxyribonucleic acid, while RNA is ribonucleic acid). What's the difference between ribose and deoxyribose? A single oxygen atom.
2. DNA forms a double helix, but RNA is a single-stranded molecule. (Sometimes that single strand will fold back on and stick to itself, though, the same way that A's/T's and C's/G's stick to each other between the two strands of DNA.) Single-strandedness is pretty important for its ability to make copies of DNA, as you'll soon see.
3. There are no thymine (T) bases in RNA. Instead, you find a different base called uracil (U). Like T's, U's are structure in a way that will stick to A's, but not to C's or G's.

RNA is used to make copies of DNA in small, transportable format – a format small enough to squeeze out through the tiny holes in the nucleus out to the general cellular mishmash outside. Once one of those RNA messages finds a ribosome, the message can be converted to an actual protein product – and stuff can finally get done. This flow of information in a cell – from stored DNA to messenger RNA to protein job-doers to actual physical traits – is known as the Central Dogma. It sadly doesn't feature a cameo from Jay and Silent Bob.

Don't Shoot the Messenger

All that fancy A-to-T and C-to-G shape matching is important to keep DNA structurally sound, but it's also the reason your cell is able to make RNA copies of DNA so easily. But what parts of DNA are worth getting copied? There are specific stretches of DNA where genes – actual information nuggets that describe how to make a particular protein – are located, and these are the ones the messenger RNAs find worth writing down. (Did you know that only about 2% of your DNA contains actual genes? There are a few bits and pieces that do other jobs, but by recent estimates, about 90% of your DNA is just there for the ride. Like I said before – DNA is lazy.)

It's actually proteins (it's always proteins) that do the job of finding the starting point for one of these genes – they look out for regions called "TATA boxes" (seriously). Technically these should be called "TATAAAAAA" boxes, because that's the sequence of nucleotides the proteins have to look out for. These TATAs are whipped out about 25 nucleotides upstream from the start of the actual gene, so once the TATA-grabbing proteins find their target, they start recruiting some of their buddies who can do the actual work of copying the gene into an RNA message.

See, proteins are good at doing jobs – but each protein is really only good at *one* job. TATA proteins are good at finding TATAs (as I'm sure many of you can empathize), but that's really the only thing they can do. All right, maybe I was hasty in calling DNA dumb. DNA is collectively dumb – as a molecule, all it can ever really do is store information. Proteins are individually dumb: they each have their own one little job, and that's all they know how to do. So TATA proteins figure out where a gene is, and throw their metaphorical molecular hands in the air to flag down somebody who can do something about that gene.

That somebody is named RNA Polymerase. Here's a helpful tip: many proteins have names ending in "-ase" that has something to do with their job in the cell. Polymer-ase makes a polymer – a polymer of RNA nucleotides, in this case. RNA Polymerase's job is to stick together a string of RNA nucleotides, but it can't just stick together any set of A's and U's and C's and G's it feels like. It works based on the template that DNA provides, in the vicinity of the gene that the TATA protein has picked out.

So RNA Polymerase sits down on the DNA strand right next to where the TATA protein is – but there's a problem. RNA Polymerase needs to check out what the DNA template says in order to make an RNA copy; but if you remember the structure of DNA, the actual information is on the inside of the molecule, where the polymerase protein can't get to it. The outside of the molecule is all phosphates and sugars; they're only there for structure and don't have any useful information that the polymerase can do anything with. And guess what? RNA Polymerase only knows how to do one job, too. Luckily, it brings some other protein friends along to help out. One of those friends is a protein that has what's called "helicase" activity. There's that "-ase" ending again, but what the heck is a "helic"? Well, it isn't anything, but apparently "helixase" sounded too weird to scientists, so they tidied the spelling up a bit. Helicase splits the bonds that hold together the A's and T's and the C's and G's in your DNA, which makes a little area of the DNA ladder pop apart – unwinding the helix in that one spot.

The whole process of finding a gene and taking down an RNA message (henceforth known as "mRNA") of it is known as transcription, and it's the first step in the Central Dogma. RNA polymerase is transcribing a copy of the DNA information, like a hard-working secretary at a microscopic office. This process is ongoing constantly with thousands of RNA polymerases at almost any given time in your cells – sometimes so frequently that multiple RNA polymerases are actually chasing one another down the same gene, and probably wishing they had tiny little horns to honk and windows to lean out of and scream.

A few things happen to a piece of mRNA once it's been transcribed, in order to make it more stable and thus more likely to live long enough to get to a ribosome and do its information-carrying job. One side effect of mRNA being single-stranded is that it's not as stable as DNA is – eventually, it'll fall apart into nucleotides, to be used again in the next round of transcription. In your cells, mRNAs get kicky little hats shoved onto their front ends (called the five-prime cap, which makes it sound like something the Borg came up with), and then something called a poly(A) tail gets stuck on to their butts – about 40 extra A's in a row streaming out behind the rest of the transcript. It's not part of the message that came from the gene – the only message it sends is a giant, 40-nucleotide-long middle finger to the proteins in your cell that roam around breaking down RNA.

With the ladder ripped open, RNA Polymerase can get in there and do its work. It makes its way up one of the separated DNA strands nucleotide by nucleotide, and whatever DNA base it finds, it adds the opposite RNA base to the strand its working on. If the DNA has a C, the RNA gets a G; if the DNA has an A, the RNA gets a U. The chain keeps growing in length – but not indefinitely. RNA Polymerase doesn't copy the whole length of the entire DNA strand starting from the TATA box – that could be millions of base pairs, and then you'd have the same nucleus-evacuation problem that you have with DNA molecules in the first place. Instead, there seem to be certain sequences in the DNA that tell the polymerase that it's not wanted any more. RNA polymerase then feels bad for itself, falls off the DNA strand, and throws away the newly built RNA strand. "I didn't really want to copy that DNA anyway," RNA polymerase tells itself. "Making RNA is stupid, anyway, and I don't even—oh hey, is that TATA protein waving at me? I should go see what he's up to over there.

(Side note: it may sound stupid to think that your cell is going to all the trouble of making these mRNAs and then also keeping chemicals around to destroy them; but keep in mind that not only does this provide a steady supply of recycled RNA nucleotides for use in the next round of transcription, but that many viruses tend to store their genetic material in the form of RNA rather than DNA. Stray, unprotected RNA floating around outside the nucleus is as likely to be a molecular terrorist as a friend, and actively destroying random RNA is the cell's hands-on equivalent to "See something, say something.")

These weird modifications serve one other purpose beyond stabilizing the baby mRNA, too. The five-prime cap in particular helps to flag down the guy whose help the mRNA needs to convert its information into an actual protein: the ribosome.

There are as many as ten million ribosomes in each of your cells, and rather inauspiciously, they are shaped like hamburger buns. There are two pieces, or subunits; one larger (the top bun) and one smaller (the bottom bun). The two pieces are floating around separately, until the bottom bun – er, subunit – finds and grabs onto a piece of mRNA in need of translating. The five-prime cap helps the mRNA stick, but the other thing is a particular sequence that will be found somewhere in the mRNA strand: the letters A-U-G in a row.

Your genes (and thus your mRNAs) are written in three-letter words called "codons", and this particular three-letter word is four-letter-wording important. AUG is called the start codon, and while there might be a few nucleotides ahead of it in the mRNA strand, this is the first one that matters. This starts what's called the "reading frame" that the ribosome will use – the thing that tells it where to start picking out three letter words. So if a mRNA sequence reads:

ACCGAUGGGUUAUUAGCAU

Then it'll be read as:

(useless crap)AUG GGU UAU UAG CAU

and not:

ACC GAU GGG UUA UUA GCA U

This is pretty important. Just imagine trying to reads omethingt hatw asw rittenw itht hew rongr eadingf rame – that's a pretty useless hunk of text, and the protein you get from reading out-of-frame is just as pointless.

So the small subunit grabs onto the mRNA, and together they grab onto the big subunit, and using the AUG codon as a guide, the ribosome begins to read along the mRNA. But how does just pulling the RNA through the ribosome do anything productive? It's not like you make anything other than a mess by pulling a spaghetti noodle through a hamburger bun, and that's basically what's going on here, isn't it?

There's one more piece to the puzzle, and that's another brand of RNA – this one is called tRNA, short for "transfer RNA". Its name comes from the fact that it actually does the work of transferring amino acids onto a growing protein. It has a funky cloverleaf shape, where its single RNA strand has bent back on itself and stuck together in places, and the two most important parts of it are at opposite ends: at the "top" of the molecule is a specific amino acid, and at the "bottom" is a three-letter RNA sequence called the anticodon – the opposite of a codon.

Remember how each letter in the RNA/DNA alphabet has its match – A's with U's and G's with C's? Well, the anticodon is the three-letter combination that matches up to the one in the codon. Each three-letter anticodon sequence is associated with one and only one amino acid – so that you get the right one of the twenty options available.

Figure 4: A tiny man dances in a blue hamburger.

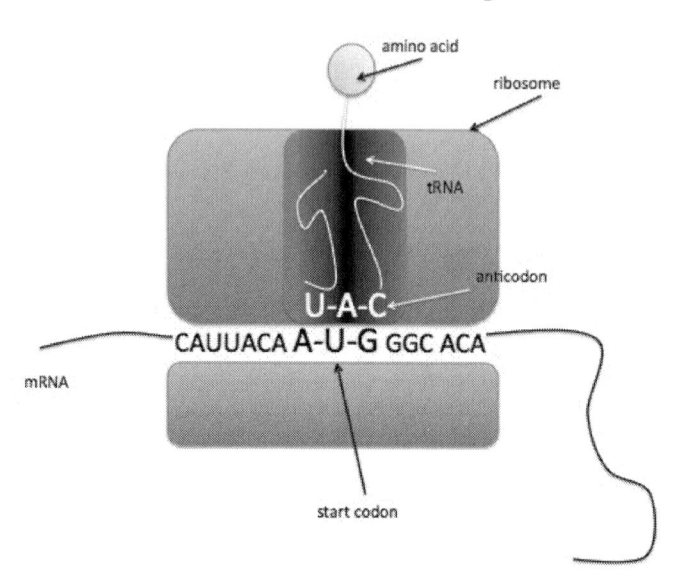

Only the tRNA with the correct anticodon to match up to the current codon available in the ribosome is allowed to drop off its amino acid for proteinification – anybody else gets quickly shown the door. (There are actually two other slots not pictured alongside the one in the diagram – one for tRNAs to enter in by, and one for them to get ejected once they've been used up by donating their amino acid to the protein in progress.)

So one by one, tRNAs that match up against the mRNA shuffle in, drop off their amino acid load, and get kicked abruptly out again. This happens over and over again, while the ribosome ratchets merrily along the mRNA strand (and in fact, several ribosomes may be scurrying down the same strand of mRNA all at once, like tiny bumper cars on a very skinny road). Each tRNA drops off a single amino acid, and each amino acid makes the protein chain that little bit longer.

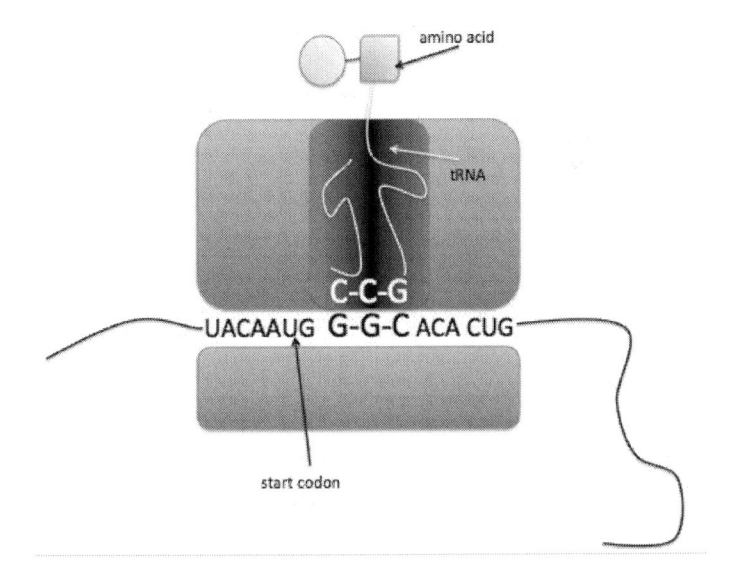

amino acid

tRNA

C-C-G
UACAAUG G-G-C ACA CUG

start codon

Figure 5: Dancing Hamburger Man 2: Electric Boogaloo

Fortunately, the ribosome never gets to the poly(A) tail, because trying to read a message of nothing but 40+ A's in a row? That's more tedious than getting 50 Facebook messages that all say "hey u" from that guy you wish realized that the "friend zone" he thinks he's in is actually the "please stop telling me that my hair looks like it would smell really good" zone. Instead, the ribosome eventually gets to one of three specific codons (UAA, UAG, or UGA) that are called "stop codons". The reason that these are called stop codons is that there isn't a tRNA that has a matching anticodon – the ribosome stops and stalls out when it finds one of these, because it can't find a tRNA to match up. Eventually, the ribosome just gives up and falls off of the mRNA.

But when the ribosome lets go of the mRNA, it also drops off the newly constructed chain of amino acids it's put together: a cute little bouncing baby protein. The sequence of amino acids helps to determine what shape that protein will have, and the shape decides what job the protein will do. If it's the right shape to stick to virus particles, it might help out in your immune system. If it happens to be built just right to reflect certain wavelengths of light, it might give you blue eyes, or brown hair. And if it's just the right shape to hold a lactose molecule, the protein might help you to not get the dairy farts after you eat a pint of Ben and Jerry's.

The information stored in DNA and transcribed in mRNA has now been translated into a functional molecule – a protein. This process, translation, is the second step in the Central Dogma (the first being transcription). If you ever forget which is which, remember that DNA and RNA speak the same language (A's, C's, G's, and T's/U's – okay, they're two different dialects, but you get the idea), while proteins speak a totally different amino-acid language of their own.

So that explains how your cells deal with biological information, which lays the groundwork for what we really need to get into to understand how DNA's behavior makes Mendel's rules possible. Different DNA sequences cause different proteins to be made – which result in the different traits old Gregor could observe. A knowledge of molecular biology also helps to explain why some of Mendel's "hereditable factors" were recessive – often, the recessive traits were those that made a nonfunctional protein. Pink flower pea plants had a perfectly good copy of the flower color gene; the plants with white flowers had mutant versions that made no protein at all. If you have one or two healthy copies, you'll get pink flowers; no healthy copies means no flower color. Short pea plants? That's a mutant version of a protein that stimulates plant growth. In this case, the protein still works – the plant grew tall enough to be short instead of nonexistent, didn't it? But it's not as good its job as the tall pea plant version.

But how does that biological information get from your parents to you? And if we each have two copies of each gene, how does it work out that our kids don't end up with four? For that matter, where do we get different "versions", or mutations, of the same genes? In short: what the heck, man?

Chapter 6: Copying the Floppy Disk of Life

Mitosis

As anyone with a computer knows (or hopefully knows), it's generally not a good idea to keep only one copy of any important file. Your cells know this basic rule of thumb (drive) too, which is why you have a complete DNA library in each of your cells. (Or most of your cells, at least – we'll outline the big exception in the next section.)

And since cells aren't immortal, you need a way to copy down all the important information in that DNA library to pass on. Your cells have a specialized process that involves doubling up on everything in the nucleus and then splitting the whole thing tidily down the middle: a complete copy for the old cell, and a complete copy for the new one too.

As we've mentioned before, DNA is stored in your nucleus in the form of chromosomes, and you have 46 of those in each of your cells – 23 from each branch of your family tree, 2 copies of each kind, as if the nucleus of your cells were some kind of Noah's Ark in miniature. When it's time for the cell to divide, each of those chromosomes has to get unzipped and copied precisely; the rungs of the ladder break in half, tearing A's from T's and C's from G's, and a molecule named DNA Polymerase (guess what that guy does) comes in and makes a new strand to pair up to each of the old ones.

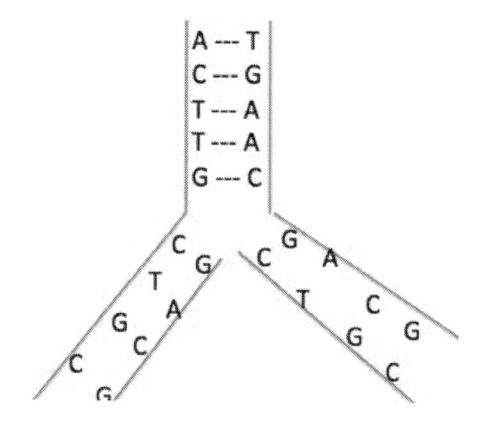

Figure 6: DNA replication fork. Not pictured: DNA replication spoon.

If you're thinking that this would take forever on a string of nucleotides millions of pieces long, you'd be right; but fortunately your cell has developed a work around. It starts splitting the double helix up and starting to grow new strands at these replication "forks" at several places along the chromosome. Then it works from each of those spots until it meets the new strands being grown from the next fork over, and joins its work together. It's a lot faster to simultaneously make a few thousand mini-strands than to make one giant one from end to end.

Before   After

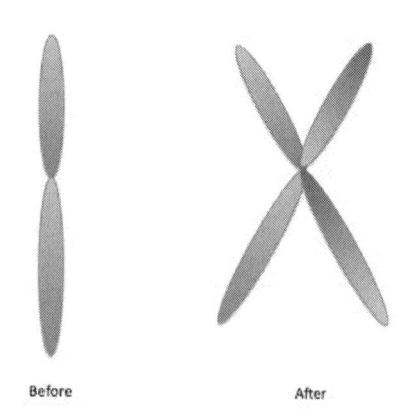

Figure 7: As you may recalled from the experiment that started all of this, DNA isn't really blue or red, but snotty whitish in color.

By the time DNA replication is over, each of the chromosomes – the wadded-up packages of DNA – has been doubled up, giving it an X-shaped appearance. Each line of the X is an individual, complete copy of the DNA in that chromosome.

Your chromosomes have been copied, but how to make sure each cell post-division is going to get a complete set? The nucleus goes through a process called "mitosis" to make sure everything gets split up even-steven.

First of all, all the X-shaped chromosomes line up down the middle of the nucleus. Then, little protein fibers grow out from one of the two opposite sides of the nucleus and attach to the middle of each chromosome . Your cell counts down ready-set-go, and those fibers start to retract. This movement snaps the duplicated chromosomes right down the middle like a molecular wishbone, and starts pulling the freshly halved chromosomes to one side or the other. Once the chromosomes are all waiting on one side of the cell or the other, like a couple of newly-picked baseball teams in PE class, the cell pinches neatly down the middle and separates them. Two daughter cells, each with a complete set of DNA instructions to send them on their merry way.

Figure 8: Yes, you have more than four chromosomes. No, I'm not drawing them.

That's how it works in most of your body's cells, which are forever in need of replenishing as you unjustly punish your liver with tequila or slough off a day's worth of skin cells. But there's one important exception to that rule – one case in which a full set of DNA is twice as much as a cell will need.

Meiosis

Mitosis makes sure that when you're making new body cells, you end up with two copies that both have one full set of DNA. But if you'll recall, when you're passing on genes to your hypothetical offspring, you don't get to pass on the full Encyclopedia of You – you only get to pass on half. So in sperm and egg cells, mitosis isn't going to cut any sort of reproductive mustard. A related process, meiosis, makes sure that the number of chromosomes gets cut in half – and cut in half in a very specific way. You can't send two copies of chromosome 1 and zero copies of chromosome 2 to one sperm cell, and zero copies of #1 and two copies of #2 to another. That would be like giving someone a copy of the Lord of the Rings trilogy that only contains three editions of The Two Towers – pointless, and with way too many pages of hobbit angst.

Meiosis starts out the same way as mitosis: your cells go through a round of DNA replication to make a copy of all your chromosomes. But wait – why bother? Why not skip that step and go straight into splitting things up halvsies? As it happens, DNA replication gives your cell a pretty good opportunity to make sure all is well in your DNA and to catch any problems before they get copied and passed on. As DNA Polymerase hops along from nucleotide to nucleotide, the DNA-copying machinery takes a moment to make sure that what it's copying makes sense. If there's an A on this strand of the old chromosome, that strand should have a T – not an A, C, or G. If there's a mismatch somewhere, the replication process gives you a chance to catch it before you inadvertently propagate the error – so going through a round of copying for the purposes of making your sex cells is a pretty good idea. Anything your body can do to avoid passing on weird mutations to the next generation helps – even if it does mean your kid has almost no chance of getting freaky X-Men powers from you.

After replication, your chromosomes once again line up down the middle of your cell for a round of team-picking, but this time, they don't line up single file. Each mom chromosome lines up side-by-side with its equivalent dad chromosome, so that when those protein fibers come in and start getting grabby, only one chromosome of each type ends up in each resulting cell. Once that's happened, the next round of division looks a lot like mitosis, except that there are only half as many chromosomes in play.

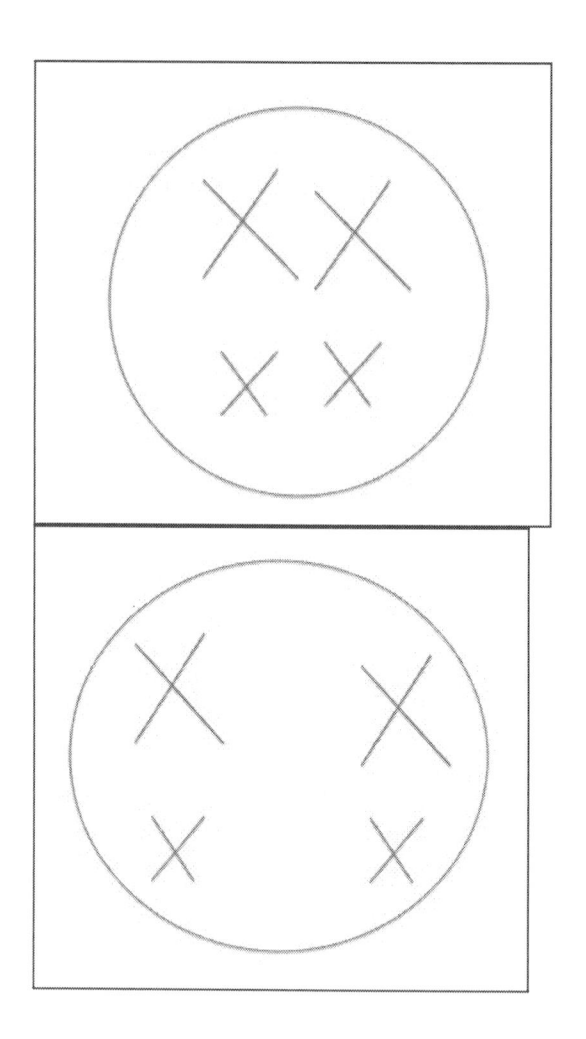

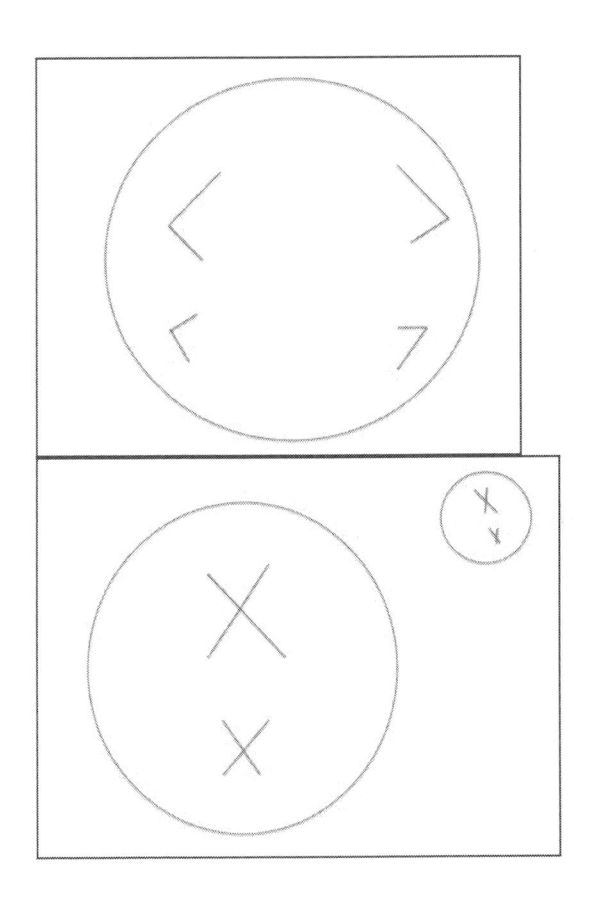

Figure 9: It's like mitosis, except more so.

So with meiosis, the cell goes through one round of DNA replication and two complete rounds of division – the number of chromosomes in the cell doubles to 92, drops back to 46 after the first division, and then down to 23 after the second. For every precursor cell you start out with, you get four little sperm cells. (In the case of eggs it gets a little more complicated – while sperm cells are streamlined, egg cells need to be ginormous in order to provide the fertilized zygote with all the cell stuff it needs to kick off its nine-month spa stay. When the first round of division happens, you get one big fat pre-egg cell packed full of cellular goodness, and one skimpily-stuffed "polar body" – a dead end cell whose only reward for making it through the next round of meiosis is to get thrown in the recycling bin. The same thing happens in the second round of division, too; the cell that's going to become the egg gets most of the goodies, and another polar body is produced just before it's un-produced. In female sex cells, a single precursor cell gives you one good egg cell and three trash-heap polar bodies. Sperm are cheap to make; eggs are expensive.)

During meiosis, the pair of matching chromosomes you got from dear old Mom and Pop get split up, but there's no rule that says any given sperm or egg cell that you make has to contain all chromosomes from Mom, or all from Dad. For each pair that you have, there's a 50-50 shot that you're going to pass on a maternal or paternal chromosome to one of your kids. Since you have 23 sets of chromosomes, the odds that you'd pass on the complete collection of either Mom or Dad's Greatest Genetic Hits to your own kid is 1 in 2^{23}, or about 1 in 8 million. This is a comforting thought – we all fear becoming our parents, let alone watching our kids turn into clones of their grandparents.

Meiosis and mitosis are annoyingly similar-sounding science words (no one ever accused scientific lingo of an excess of creativity), so just to recap:

Only meiosis	Both mitosis and meiosis	Only mitosis
* Two rounds of division	* One round of DNA replication	* One round of division
* Halves number of chromosomes		* Maintains number of chromosomes

* Used to make sex cells		* Used to make literally any other kind of body cell
* Wears short skirts		* Wears t-shirts
* Is cheer captain		* Is on the bleachers

Once a sperm or egg cell has been produced in your genital region of choice, they can be smashed together to reconstitute that original 46 chromosomes that everyone started out with. The fact that you have two copies of each chromosome ties back to those rules that Mendel figured out – each of Mendel's "hereditable factors", or genes, has an actual physical location on one of those chromosomes. You get one copy from each parent, pass one copy to any kid that you may have, and the carousel of life goes around one more cheery turn.

Unless, of course, something goes wrong along the way. Cue the ominous organ music!

Chapter 7: Genetic Health Goes "Boink"

Where's the Laser Vision Gene?

Your body is a delicate, complicated machine, and it's no less likely than any other piece of complex technology to randomly stuff a brick. Ever had your iPod go kerflutz for no apparent reason? Ever had your modem give up and die in the middle of a YouTube video just because it apparently lost the will to go on? Yeah, your cells do that too, sometimes. The good news is that while you probably only have one modem, you have billions and billions of cells to pick up the slack – for the most part.

Let's start with where things can go horribly awry at the lowest level of detail: the actual DNA sequence itself. If you have ever seen a comic book in your life, you have probably heard the word "mutation"; if you have associated that word with "awesome superpowers", though, it's time to disabuse yourself of that notion. Mutations can be caused by any number of things (UV light exposure, radiation, that blackened stuff on an overcooked hot dog), but however they happen, they work pretty much the same way.

There are three "styles" of mutation. One is called "neutral" – it doesn't have any noticeable effect at all. Maybe it happens in one of the vast swaths of DNA that don't code for any genes; or maybe the change just doesn't matter. Remember how there are twenty different amino acids that make up proteins? And how it's a three-letter combination in your DNA that keys in the code for which amino acid to use? And how there are four different letters to use in that combination? I know that's a lot of remembering, but bear with me.

There are four possible letters to put in each of the three slots in a codon. Four times four times four (4^3) = 64 possible combos. 64 (I'm going to get kind of math-y here, hang on) … is more than 20. That means your DNA code has some redundancy built in – for certain amino acids, more than one combination can specify the same amino acid. For example, the codons GUU, GUA, GUG, and GUC are all valid ways to ask for the amino acid called valine. This is called the "degeneracy" of the genetic code, which makes it sound like the code is up to no good, but really, things had to work out this way. If codons were only 2 letters long instead of 3, you could only have 4^2 (16) possible "words" in the genetic dictionary, and that's not enough to spell out 20 different amino acids. Given the choice between 16 and 64 words, evolution went with 64, which just means there's a little flexibility in the system.

So if you get one of the mutations that just swaps a GUU codon for a GUA, no big deal – you're still going to end up with the same amino acid in the finished protein product. No information lost, none gained. In some cases, where a mutation does actually change the amino acid, there still might not be any real effect – certain amino acids are shaped a lot like, and so function a lot like, others, and swapping them around might not matter very much. Neutral mutations are the most common type of mutation; most of the time, instead of giving you the ability to control the weather, mutations just don't do anything at all.

The next most common mutation style is the harmful kind. These are mutations that turn off genes that you want to have turned on, or vice versa; or that substitute in completely wrong amino acids that wreck the shape of a protein. Imagine a mutation that turns the TATA box into the GATA box – now that TATA protein has no idea where to grab on to get an mRNA copy of that gene made. Or how about a mutation that turns an UGG codon into a UGA codon? UGG codes for the amino acid tryptophan, but UGA is a stop codon. Now you've cut off making your protein too early, and half a protein is generally not much better than no protein at all.

Finally, and most rarely, mutations can actually be helpful. A change in amino acid sequence could actually make a protein better at doing its job – helping it bind more tightly to the sugar it breaks down, making it extra durable in the face of dangerous chemicals, etc. (Note that "helpful" still does not mean "metal exoskeleton". Sorry.)

Besides talking about mutations in terms of their effect on the body, we can also talk about what exactly they look like. There are a few broad categories of mutation: substitutions, deletions, and insertions.

Substitutions are exactly what it says on the label: one nucleotide in the sequence gets swapped out for another. If the sequence used to say ACCG, maybe now it says TCCG. Oops! Substitutions can be pretty harmless (see the above example involving two equivalent valine amino acid codons), or they can cause pretty big problems – a mutation in the TATA box will stop a gene from being transcribed in the first place, and a substitution that makes a stop codon where there didn't used to be one will still get you a protein, but a stupid, useless protein.

Insertions and deletions are bigger problems. Remember how DNA is written in three-letter codons that give the ribosome a reading frame in which to match up amino acids? Guess what happens to that reading frame when you add in a single nucleotide, or take one out?

A sequence that originally read:

AUG CCC GGC UAU CCG ➔ met – pro – gly – tyr – pro

With an insertion becomes:

AUG CCC CGG CUA UCC G ➔ met – pro – *arg – leu – ser*

And with a deletion becomes:

AUG _CCG GCU AUC CG ➔ met – pro – *ala – ile*

After an insertion or deletion, every single codon downstream of the problem is going to be out of frame, and it's going to get you the wrong amino acid. How well do you think that protein's going to turn out? Try making a pizza, but halfway through, start substituting in a completely different ingredient for everything listed. "Yeah, instead of cheese, we thought we'd use frozen yogurt. How do you like it?"

If you're lucky, your cell will figure out something is amiss and hit the self-destruct button – a process called "apoptosis" in which the cell basically eats itself in a miniature ouroboros of destruction. That way, the mutation is nipped in the bud, and the cell that contained it is killed off before it can reproduce and carry the error on down the line. Otherwise, mutations can start to stack up – and too many of those in the wrong places are how tumors can arise.

Mutations in the sex cells, on the other hand, might not cause any problems for you personally, but it could cause some issues for your unfortunate offspring. There was no history of hemophilia in the British royal family until Queen Victoria's kids hit the scene – scientific historians suspect that one of Her Royal Vickiness's precursor egg cells had a mutation in a gene that affects blood clotting. Hemophilia has been plaguing the noble houses of Europe ever since (or lycanthropy, depending on if you get your historical information from primary sources or from Doctor Who).

Mutations in specific genes, whether freshly mutated or passed unwittingly by a carrier parent, can cause all kinds of diseases – hemophilia, cystic fibrosis, sickle cell anemia, and more. But what happens when the errors are on a bigger scale – a change to one gene is bad, but what if Bad Stuff happens to a whole lot of genes at once?

My, Oh, Meiosis

Your cell has a well-organized method of lining chromosomes up during mitosis and meiosis and then splitting them down the middle, which always runs like clockwork – except when it doesn't. Every now and again, those protein fibers will miss their target chromosome, and wind up grabbing the wrong one instead. This means that instead of neatly dividing things in half, you wind up with one cell that has an extra copy of a particular chromosome, and one cell that gets shorted. This is called a "nondisjunction", which is science-talk for "someone's gonna have to pay for that", and neither cell is particularly happy with the results.

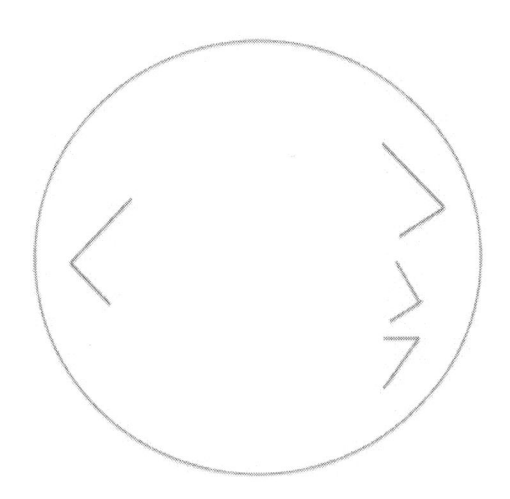

*Figure 10:
Somebody
done goofed.*

If this happens in a sperm or egg cell, that cell then meets up with the opposite cell variety to make a mini-me, the situation is seriously suboptimal. If the zygote gets shorted on a chromosome (a situation called "monosomy"; *mono-* for one and *–somy* for, you know, chromosome stuff), it's not going to have spare copies of any of those genes. If there are any crummy alleles for genes on the single chromosome copy that it does get from the unaffected parents, there's no backup to rely on – and that can be a problem for hundreds of different genes. In some cases, a zygote that's monosomatic for a particular chromosome can't even survive – for example, a Y sex chromosome without an X chromosome to provide backup is fatal (although an X without a second X or a Y is okay).

So having an extra copy just means an extra backup, right? All should be well for people with trisomy – three copies of a particular sort of chromosome? Well ... not exactly. The human body is fine-tuned to make do with at most two copies of a lot of genes – having a third copy around getting busy dictating what proteins to make can result in cells basically "overdosing" on those proteins. Down Syndrome, Edwards Syndrome, and Klinefelter's Syndrome are all examples of human trisomy, and are, to a varying degree, not much fun.

Besides chromosomes failing to fall into line appropriately, other weird things can happen when the cell is hurriedly trying to package things up for division. Huge pieces of chromosomes can break off and be lost (deletions); they can snap off the end of one chromosome and find themselves stuck to another (translocations); or a stretch of DNA can be inadvertently duplicated, making its chromosome a bit longer than it used to be. The consequences are obvious in the case of a deletion event – bye bye, genes; and translocations are also pretty apparently bad, if you snap a chromosome in half right in the middle of a gene – it's not as if you can make two different mRNAs and then glue the resulting proteins together after the fact. But where's the harm in a duplication?

If you've ever heard of Huntington's Disease, you'll have an idea of what the problem can look like. Huntingtin (with an I instead of an O) is the name of the protein involved in the disease – it seems to show up in most body cells, and plays an unknown role in the brain and development. In the gene that codes for the protein, there's a certain stretch that contains the same three nucleotides, repeated in a pattern over and over: CAGCAGCAGCAG. The number of repeats varies from person to person – you might have five or six or a dozen repeats, and you'd be fine. But the problem is that DNA Polymerase is not particularly good at counting, and sometimes, it loses track while copying the Huntingtin gene. "How many CAG's have I put together already? I'll just stick in a few more. Wait, did I do the CAG's yet? Better safe than sorry."

Up to 25 CAG repeats and you won't have any problem; between 25 and 35 or so things are a bit iffy; and over 35 means you have a serious problem. Not only is the protein severely misfolded with an extra three dozen amino acids wedged into its middle; but its funky shape makes it "sticky" its mutant-Huntingtin comrades. The misformed proteins clump together and can kill the cell they're in, turning the region into the bodily version of Swiss cheese – wait, which cells does Huntingtin play a major role in? Oh, good, the brain. Huntington's disease is biology's way of reminding us that too much of a good thing – in this case, a gene necessary for the body's development – is very much not a good thing.

Chapter 8: Exceptions That Prove the Rule

It's Never that Simple (Except When It Is)

If up to this point, you've been astonished that a monk in the 1800s could uncover a complete understanding of heredity, don't worry: Mendel laid the groundwork, but things get a lot more complicated than "yellow or green peas", especially when it comes to human genetics. (Or Adder's Tongue ferns, which have about 1200 chromosomes; or sleeping sickness parasites, which randomly – or not so randomly – add extra U's all throughout their mRNAs once they've been transcribed, or—okay, so maybe human genetics isn't anything super special, but it does hit a little closer to home.)

There are many cases where Mendel's rules just aren't enough to explain what's going on, because what's going on is way cooler and more interesting than dominant-versus-recessive. Rarely is any trait controlled by a single gene that has only one allele, one of which is 100% dominant and the other of which is 100% recessive. Ever gotten a suntan (or, if you're like me, a sunburn)? Congratulations – you've just modified the expression of your genes using your environment. Have an eye color outside of blue or brown? Ooh – it turns out things are a little more complicated there than they told you in biology class, too. Sorry, Greggers, but things just aren't that black and white (or pink and white, maybe) for most real-world traits.

Incomplete Dominance

As previously discussed, if you have two different alleles of a gene, the dominant should hide the recessive one. But remember that the dominant gene is actually doing that hiding by way of a protein product. What if the amount of protein you can make from one copy of a gene just isn't enough to keep up with all of a cell's needs? In that case, you get incomplete dominance. An example is flower color in certain (non-pea) flowering plants.

In snapdragons, there are two different flower color alleles: red, and white. (The "white" allele is actually technically a "colorless" allele – the mutation. If you get two red alleles, you get a red-flowered snapdragon. If you get two white alleles, you get white flowers. But if you get a red and a white allele, the resulting flowers are pink – not white. The snapdragon's cells just can't make enough red pigment to color the whole flower red (no thanks to that lazy-ass white allele), so we can't really say that red is fully dominant to white – just incompletely so.

Codominance

This sounds like incomplete dominance, but it's not the same – I promise! In incomplete dominance, the trait is being expressed in a half-assed way that's neither really the dominant version nor the recessive one, but sort of a blend between the two. In the case of codominance, though, you get two distinct versions of the trait both being expressed at the same time. The obvious example here is human blood typing.

Your blood type is either Type A, Type B, Type AB, or Type O – if you have a blood donor card, you may know which one. (People used to do an experiment to find out in biology classes, before one too many students, after comparing their blood types to their parents', accidentally figured out that they were adopted.) There's one gene that determines which one you've got, and three different possible alleles for that gene: A, B, and O.

If you've got two O alleles, congratulations – you have the most common blood type (which is good, coincidentally, because you can't get blood from donors of any other blood type – but they can all get blood from you). O, however, is recessive to either A or B alleles, so if you have an A and an O, you're type A, and there are little A-shaped molecules all over the outside of your red blood cells. (O is recessive because there are no O-shaped molecules; type O people have basically naked blood cells.) If you have a B and an O, your blood cells are covered in B-shaped thingers and you have type B blood (and you have to endure a lot of "BO" jokes).

But if you get an A and a B, you have both A-shaped molecules and B-shaped molecules on the surfaces of your blood cells. There's no weird A-B hybrid molecule – there are still two distinct alleles, making two distinct proteins that make two different cell surface doodads. They both show up and neither one is hidden – the A and B alleles are co-dominant.

Pleiotropy and Epistasis (Are Not Two Words I Just Made Up)

In certain cases, a single gene can have an effect on more than one trait. For example, there's a gene in chickens called "frizzle" (not, sadly, after the Magic School Bus teacher); a certain mutation in it causes chicken feathers to curl outwards instead of lying tidy and flat – to frizzle, if you will. But this same gene has other effects on the chicken too – the same mutation is associated with these chickens being better able to cope with hot temperatures, with a rapid metabolism that has them burning through chicken feed like it's going out of style, and with low rates of egg-laying. They look pretty fancy, but frizzled chickens are not exactly the rage among egg farmers.

Epistasis is something of the opposite of pleiotropy. In pleiotropy, one gene affects multiple traits; in epistasis, one gene controls the expression of another, so that two genes are necessary to create a single trait. Albinism is a famous example of this particular genre of non-Mendelian shenanigans. There are eye and hair and skin color genes that make the pigments that give these body parts their particular hues, but there's also a gene that makes a transporter molecule to actually move those pigments to where they're supposed to be. You can have a perfectly good brown eye color gene, and make all the brown eye pigment you want – but without a functional copy of the gene that codes for pigment-transporting protein, that brown pigment isn't going anywhere. (The red coloring in the eyes of albino people is from the blood vessels there – those are in everyone's eyes, but usually there's some kind of pigment there to mask them.)

Polygenic Traits

Unlike pea plants, humans do not come in discrete "short" or "tall" varieties. We also don't have only two specific skin colors, or distinct "smart" and "idiotic" groupings (the comments on YouTube being the obvious exception). What allows for the gradual continuum of heights, skin tones, intelligence, and other such traits?

Polygenic traits – traits with several genes contributing to their expression – typically vary along a wide and gradual range. You only need one gene to decide whether or not your tongue rolls, but shockingly enough, most other facets of the human condition get a little more complicated than that.

How do we get this bell curve of options for these traits, though? Let's take the example of skin color. We can simplify and say there are four different genes that affect skin color, and for each of them, you have an either-or option – either a mega-melanin allele, or a mayonnaise-and-wonder-bread allele. Chances aren't great that any given person is going to wind up with two mayos for every single one of those four genes – but a few people are going to, and they make up one tail of the bell curve, while the other end is made up of people who wound up doubled up on melanin-making alleles in each of the eight slots available. Most people are going to wind up somewhere in the middle of the bell curve – four or five or six darker alleles mixed in with two or three or four light ones.

So there you have it – the wonderful world of genetics, which is neither as complicated as it looks from the outside, nor as simple as it looks from the purview of a nineteenth-century pea gardener. Hopefully you know enough now to pass your biology quizzes, or to read an issue of *Scientific American* without getting the flop sweats. And if you're still interested in knowing more, go read Matt Ridley's book *Genome*; he got a lot more Smart Guy alleles than I did, and more than a few extra Good Writer alleles, too.

Printed in Great Britain
by Amazon